WHEN DINOSAURS RULED THE EARTH

THE TYRANNOSAURUS REX

Written by Tracy Vonder Brink

Illustrated by Riley Stark

TABLE OF CONTENTS

A Crabtree Seedlings Book

Crabtree Publishing

crabtreebooks.com

School-to-Home Support for Caregivers and Teachers

This book helps children grow by letting them practice reading. Here are a few guiding questions to help the reader with building his or her comprehension skills. Possible answers appear here in red.

Before Reading:

• What do I think this book is about?
  • *I think this book is about dinosaurs.*
  • *I think this book is about the dinosaur called* Tyrannosaurus rex.

• What do I want to learn about this topic?
  • *I want to learn how big* Tyrannosaurus rex *was.*
  • *I want to learn what* Tyrannosaurus rex *ate.*

During Reading:

• I wonder why…
  • *I wonder why paleontologists study fossils.*
  • *I wonder how* Tyrannosaurus rex *used its tail.*

• What have I learned so far?
  • *I have learned that dinosaurs lived before people.*
  • *I have learned that* Tyrannosaurus rex *lived in North America.*

After Reading:

• What details did I learn about this topic?
  • *I have learned that* Tyrannosaurus rex *had three toes on each foot.*
  • *I have learned that* Tyrannosaurus rex *had 60 teeth.*

• Read the book again and look for the glossary words.
  • *I see the word* **paleontologists** *on page 4 and the word* **carnivore** *on page 17. The other glossary words are found on page 22.*

TYRANNOSAURUS REX

Many dinosaurs once roamed Earth.

They lived long before people.

Some dinosaurs became **fossils** after they died.

Paleontologists study fossils to learn about dinosaurs.

Tyrannosaurus rex was a dinosaur that lived about 67 million years ago.

Tyrannosaurus rex fossils have mainly been found in North America.

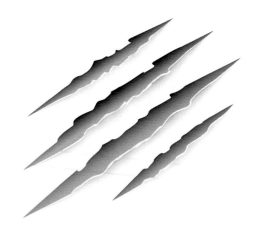

Area where *Tyrannosaurus rex* fossils found

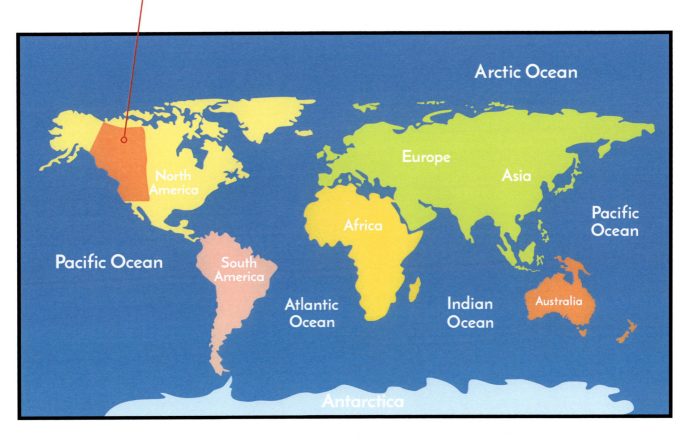

Tyrannosaurus rex *fossils have been found in the United States and Canada.*

Tyrannosaurus rex grew up to 40 feet (12 m) long.

That's as long as some school buses.

Tyrannosaurus rex walked on two legs.

Each foot had three toes.

Tyrannosaurus rex *may have been able to run as fast as a human.*

Tyrannosaurus rex had a large tail.

Its tail helped it **balance** while walking and running.

Tyrannosaurus rex had two short arms.

Its arms were much stronger than a human's.

Tyrannosaurus rex *was one of the largest meat-eating dinosaurs.*

Tyrannosaurus rex was a **carnivore**.

It was a large **predator** that hunted other dinosaurs.

Tyrannosaurus rex had 60 sharp teeth that could crunch through bone.

Did it hunt alone or in a **pack**?

So far no Tyrannosaurus rex *fossils have been found grouped together in a pack.*

19

How did *Tyrannosaurus rex* use its arms?

More fossils need to be discovered to find the answers.

This Tyrannosaurus rex *skeleton is nicknamed Sue in honor of Sue Hendrickson, the fossil hunter who discovered it.*

Glossary

balance (BAL-uhns): To keep the body steady without falling down

carnivore (KAR-nih-vor): An animal that only eats meat

fossil (FAWS-uhl): The traces, prints, or remains of plants and animals that lived long ago

pack (PAK): A group of animals that hunt together

paleontologist (PAY-lee-en-TAW-luh-jist): A scientist who studies fossils to learn about past life on Earth

predator (PREH-duh-ter): An animal that hunts and eats other animals

Index

About the Author

Tracy Vonder Brink loves to learn about science and nature. She has written many nonfiction books for kids and is a contributing editor for three children's science magazines. Tracy lives in Cincinnati, Ohio, with her husband, two daughters, and two rescue dogs. Her favorite *Tyrannosaurus rex* is Sue, at the Field Museum in Chicago.

Written by: Tracy Vonder Brink
Illustrated by: Riley Stark
Designed by: Rhea Wallace
Series Development: James Earley
Proofreader: Melissa Boyce
Educational Consultant: Marie Lemke M.Ed.

Photographs:
Shutterstock: ShutterStock Studio: p. 5; Rimma R: p. 7;
    PaoW: p. 21

Crabtree Publishing

crabtreebooks.com   800-387-7650

Copyright © 2024 Crabtree Publishing

Printed in the U.S.A./022024/PP20240115

Published in Canada
Crabtree Publishing
616 Welland Ave.
St. Catharines, Ontario
L2M 5V6

Published in the United States
Crabtree Publishing
347 Fifth Ave
Suite 1402-145
New York, NY 10016

Library and Archives Canada Cataloguing in Publication
Available at Library and Archives Canada

Library of Congress Cataloging-in-Publication Data
Available at the Library of Congress

Hardcover: 978-1-0396-9646-4
Paperback: 978-1-0396-9753-9
Ebook (pdf): 978-1-0396-9967-0
Epub: 978-1-0396-9860-4